STRAWBERRY
DEFICIENCY SYMPTOMS:
A Visual and Plant Analysis Guide to Fertilization

Albert Ulrich, M.A.E. Mostafa, and William W. Allen

Agricultural Experiment Station
UNIVERSITY OF CALIFORNIA
Division of Agriculture and Natural Resources

The authors:

Albert Ulrich is Plant Physiologist Emeritus, Department of Soils and Plant Nutrition, Berkeley.

M. A. E. Mostafa, deceased, was a Staff Research Associate, Department of Soils and Plant Nutrition, Berkeley.

William W. Allen is Entomologist and Lecturer, Department of Entomology and Parasitology, Berkeley.

For ordering information and a free catalog of current publications, write to:

Publications
Division of Agriculture and Natural Resources
University of California
1301 S 46th Street
Building 478-MC 3580
Richmond, CA 94804

Telephone (510) 665-2195   FAX (510) 665-3427

POD 09/10

ISBN 0-931876-37-0
Library of Congress catalog card number 79-67379

TABLE OF CONTENTS

THE FABULOUS FRUIT

No matter where it grows—near the seacoast, in forests, in gardens, or in commercial fields—the strawberry plant is cherished for its beautiful red fruit that has a tantalizing aroma and delicious flavor. As early as 1360, the famous king of France, Charles the Fifth, carefully selected plants he found in the woods to plant in his gardens for the enjoyment of his royal friends. By 1386 he had as many as 12,000 plants. Much of the romance with wild strawberries was recorded in early religious paintings and can still be seen in the museums of Europe. Fruit from descendents of such "wild" strawberries, specially grown, may still be purchased in Europe's markets.

In the New World, species were discovered in Canada, Virginia, and Chile. When they were introduced in Europe, a new, larger, still more tantalizing strawberry was accidentally hybridized. These hybrids were returned to the New World, where they thrived in commercial fields near the large city markets. Still later, a selection from the native California beach strawberry with its long-bearing habit, came to play its role.

But the latest and perhaps most dramatic chapter in the strawberry story, took place in California in the mid-1940's. That's when geneticists and horticulturists from the University of California introduced five new market varieties—two of which, the Shasta and the Lassen, were to expand this state's strawberry crop, already worth $12 million annually, to one worth over $30 million annually in ten years.

The success story continued. Pomologists developed even more productive varieties of high quality; plant pathologists discovered that major plant diseases could be controlled by soil fumigation and by planting virus-free certified nursery plants; entomologists devised effective pest control systems. And finally growers, through experience, grew strawberries wherever sunlight, temperature, and soil fertility were suitable for best plant growth and fruit production. A unique system was also developed to assure orderly marketing of high quality fruit at the consumer's favorite fruit market.

At this time the strawberry industry is worth $400 million annually. Yields of only a few tons per acre common 50 years ago have risen to as much as 50 tons per acre, and the acreage has increased from 5,000 acres to 21,000 acres in 1991.

Now with the strawberry as one of the most popular dessert crops, the fabulous fruit is flown from California to appreciative connoisseurs all over the world to whom it is served in simple, unadorned elegance, or delectably prepared in innumerable and imaginative ways, including with whipped cream and cake—or even with champagne, sparkling in soft candle light.

And its fertilizer needs

In growing strawberries, either as a hobby or for the market, high yields of quality fruit depend on proper cultural practices, including the judicious use of fertilizers. No two fields are exactly alike, and because soil type, climate, cultural practices, and even varietal characteristics can differ so widely, no blanket fertilizer recommendation can be made to include every situation.

In most instances, a grower will stay with his fertilizer program as long as yields and quality of fruit are satisfactory, and fertilizer costs are manageable. Sometimes, however, he believes he can improve his fertilizer practices and frequently begins to play a guess-and-guess again game that leads more often than not to over-fertilization.

Over-fertilization, especially with nitrogen, leads to excessive leaf growth, poor fruit quality, and sometimes to albinism, particularly in susceptible varieties. A slight deficiency of nitrogen, on the other hand, lowers yield somewhat but beneficially reduces or eliminates albinism and often enhances fruit sweetness, fruitiness, and firmness.

So what is the answer? Clearly, nitrogen should be used moderately, and the grower must keep a watchful eye for all nutrient deficiency symptoms. Asking the plant about its nutritional needs is the best way for the grower to use fertilizers—especially nitrogen—more effectively on strawberries. This is accomplished by first visually inspecting and diagnosing foliage and fruit, confirming the visual diagnosis by testing plants chemically, and then applying fertilizer. Ideally, the grower or technical advisor should collect and analyze plant samples systematically in a plant analysis program, with the results serving as a guide to the fertilization of each "patch" or block of strawberries. (See Figure 1.) Without a continuous supply of nitrogen, yields decrease dramatically (Figure 5, page 49 and Table 2, page 47).

There are essentially three methods currently available to evaluate and improve a fertilizer program: 1) visual examination and identification of deficiency symptoms, 2) field and/or pot tests, and 3) plant and soil testing. Whatever the method chosen, success will depend on good grower skills, local technical services, and correct interpretation of results.

The three methods differ greatly in providing information on the fertilizer needs of strawberries. Decisions based on visual inspection assume that specific symptoms (as presented here by photos) characterize a particular nutrient deficiency. This is a quick and easy method to be used by an experienced grower; unfortunately, much damage to a crop can have already taken place by the time the deficiency becomes visible. The second method, diagnosis based on field and pot testing, is accurate but time-consuming and costly.

The third method, plant and soil testing, is rapid and far less costly but the tests are best calibrated against known plant growth responses—as shown for plant tissue tests—(Figure 1). In the case of the soil test, it is assumed that what is in the soil is somehow related to plant growth, even though the soil is at least one step removed from the plant. As a rule, soil test values can be interpreted correctly when the values are either very low or very high. Intermediate values, however, are difficult to interpret because climatic factors and cultural practices strongly affect both the nutritional supplies from the soil and the nutrient needs of the plant for growth and development.

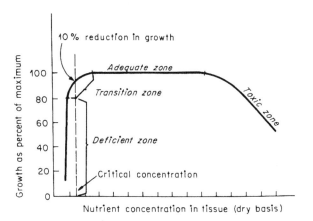

Fig. 1. Growth of a crop related to the concentration of a nutrient in the tissue. The critical concentration is usually taken at the point where growth is 10 percent less than the maximum. Deficiency symptoms usually appear below the critical concentration and not at all above it. The sharper the transition zone, and the broader the range of values from deficiency to toxicity, the more useful the calibration curve becomes for diagnostic purposes. Toxicity symptoms may or may not appear in the toxic zone as a nutrient becomes excessive. When toxic symptoms do appear, stage of growth and components of climate, particularly air temperatures and wind velocity, alter symptoms greatly.

In contrast to soil tests, Plant tissue tests measure directly what is in the plant and are directly related to plant growth. When the nutrient concentration within the plant decreases below the critical level, the growth rate, and therefore, the yield, also decrease (Figure 1). Plants below the critical level will respond to fertilization. Naturally, the earlier in the growing season a deficiency is detected, and the sooner it is corrected, the greater the response to fertilization.

IDENTIFICATION
OF NUTRIENT DEFICIENCIES

Descriptions of Visible Symptoms

Symptoms on Plants are signs of stress and are related to specific causes. Wilting, for example, is a characteristic symptom of water shortage caused directly by a lack of water or indirectly by root damage from cultivation, disease or pests, or by excessive leaf transpiration. Similarly, various leaf symptoms are caused by specific nutrient deficiencies or by toxic substances. Systematic classification of symptoms, confirmed by chemical analysis of the affected leaves, should therefore provide a quick and easy way to determine what nutrient deficiency or toxicity, if any, is the cause of the plant's stress symptoms. See Key (opposite) and Table 1 (page 42) of analytical values for a rapid comparison of deficient and non-deficient leaves.

The symptoms discussed here are the ones used in the Key and in the Color Atlas, pages 14 to 41. The grower who can learn to recognize these symptoms and their possible causes has taken the best preliminary step toward correcting a possible nutritional problem.

LEAF SYMPTOMS

Uniform yellowing   The strawberry plant in its natural habitat is remarkably free of mineral deficiency symptoms, or so it may appear to the casual observer who is about to pick a luscious red berry from a wild strawberry plant in the woods or near the sea coast. On the other hand, the close observer might see leaves ranging in color from light-green to golden-yellow and, at times, splashed with red. The leaf stalks (petioles) will often be reddish—nearly always a sign of nitrogen deficiency. The diphenylamine

Stunted/dark-greening   Dark-green plants, growing rather slowly without clearly defined symptoms other than small leaf size, or occasionally with a purpling of the underleaf surface are most likely deficient in phosphorus. A low 2 percent acetic

Leaf scorch/salt burn   Leaf scorch may be due to potassium or magnesium deficiencies. However, it can also be caused by wind burn, a temporary shortage of water, excessive salts or several plant diseases. In these latter instances, leaf-blade analyses will

Growing-point damage (tip-burn)   Growing-point damage, associated with deformed fruit and poor root development, indicates a deficiency of calcium or boron. Leaf symptoms of both are similar during the early stages of development. Both cause puckering and tip-burn of the younger leaves, followed by marginal yellowing and

Yellowing with green veining   Yellowing with green veining indicates zinc,

FRUIT SYMPTOMS

Fruit may be deformed because of poor pollination associated with environmental, mechanical, chemical, pathological, entomological, or nutritional injury. High temperature, for example, prevents pollen formation, and fruit fails to form. Injury may appear in association with nutritional excesses or deficiencies. A deficiency of sugar, for example, induces albinism, whereas, a deficiency of boron causes bumpy fruit and that of calcium, hard fruit. A deficiency of potassium is associated with insipid, flavorless fruit,

Poor pollination   Boron deficiency greatly reduces pollination so that the few pollinated areas have a raised, bumpy appearance. High and low temperatures also

Hard fruit   Calcium deficiency prevents normal fruit enlargement without reducing pollination. Local areas on the fruit have high densities of seeds (achenes) per

Insipid fruit   Potassium deficiency causes fruit to be insipid, colorless, pulpy, and soft in texture. Such fruit is often associated with leaf scorch and potassium defi-

Albino fruit   Albinism, or white fruit disease, results in soft, puffy, tasteless fruit that spoils rapidly before it ripens. The achenes of albino fruit are sunken and closely surrounded by red and then puffy white tissue, which turns from white to yellow, and eventually to brown. Generally, albino fruit appears in susceptible varieties when plants

DEFICIENCIES

Key to Color Atlas

If your strawberry plant shows symptom (left below), compare it with photos on pages listed opposite.

analytical tests (Table 1) can be used to determine nitrogen, sulfur, or molybdenum deficiencies. If tests indicate these nutrients are above deficiency ranges, yellowing is caused by poor soil drainage, low air or soil temperature, plant disease, or the like.

acid soluble H_2PO_4-P or total-P content in leaves and/or a response to phosphorus fertilization will confirm the diagnosis.

show that potassium and magnesium are above deficiency ranges.

crinkling, and reduced growth of the growing point. In time, however, deficiency symptoms can be distinguished from one another, as can be seen in the Color Atlas.

manganese, iron, or copper deficiency.

whereas, a deficiency of nitrogen enhances sweetness and flavor. Modifications by various deficiencies, especially in relation to variety, are yet to be studied.

may prevent pollen formation.

unit area, and the flesh has a "hard" texture.

ciency in leaf blades.

are bearing heavily during rapid fruit enlargement and when they become low in sugar because of reduced photosynthesis during cloudy weather.

Strawberries growing in the greenhouse by the nutrient solution culture technique for inducing nutrient deficiency symptoms.

COLOR ATLAS

The corrective measures suggested in this section are general recommendations. For local experience and specific details on fertilization and cultural practices, see your farm advisor or technical consultant.

Nitrogen is the key element...

Field experiment at San Jose, California. Small plants, nitrogen-deficient; large plants, fertilized with nitrogen.

Nitrogen-deficient Tioga plant grown in solution culture.

Smaller, normal-appearing flower from nitrogen-deficient Tioga plant (left), larger flower from nitrogen-sufficient plant (right).

(For more on nitrogen, turn page.)

Symptoms vary from slight to conspicuous depending on leaf age and the degree of deficiency. At first, especially during periods of rapid growth, leaves gradually turn from a characteristic green to a barely distinguishable light-green. As deficiency progresses, the leaves become uniformly yellow and smaller. If the diphenylamine test (page 42) for nitrate is positive, or the petiole and analytical values are appreciably more than 500 ppm nitrate-N (dry-basis), the yellowing probably is due to a shortage of sulfur or molybdenum or poor drainage or some other cause.

Supply of nitrogen from the soil and environmental conditions will determine the degree of deficiency and appearance of symptoms. With a slight deficiency, the plants will appear almost normal, especially on heavy soils with nitrifiable organic nitrogen. Here, the older leaves will be light-green and somewhat smaller than normal. Frequently, as leaves age, leaf stalks and calyxes (caps) become reddish, and leaf blades become a brilliant red. In contrast, young immature leaves become greener with increased nitrogen deficiency. Fruit size becomes smaller as the deficiency becomes greater. However, a slight nitrogen deficiency may be desirable, for it improves fruit quality and, despite a slight loss in yield, brings a tastier berry.

Plants deficient in nitrogen will often appear to have recovered under field conditions. New leaves form and develop in size commensurate with the rate of nitrate release from soil organic matter through nitrification. New nitrate is absorbed by the fibrous roots and is readily elaborated to amino acids and amides, or it moves to the young leaves to be utilized there. Under these conditions leaf blades become green and, except for their smaller size, appear normal. A young, fully expanded leaf of this type will give a negative petiole test for nitrate with diphenylamine reagent. Fields with green leaves such as these escape detection, unless the plants are tested periodically with diphenylamine reagent in the field or analyzed more precisely in the laboratory.

Corrective measures

Application of readily available N to the soil as a side dressing at the rate of 56 kg / ha (50 lb N / acre) equivalent to 6.5 g / m (6.5 oz / 100 ft row), followed by an irrigation, or alternatively, through the drip or soaker irrigation system, will correct the deficiency. (See also page 55.)

Close-up of nitrogen-deficient plants (right) with nitrogen fertilizer (left). Tioga, upper row; Tufts, lower row. San Jose, California.

Leaves from a nitrogen-deficient Tioga plant (young to old). Youngest blades (upper left) are much greener than mature blades (lower left). Oldest blades are light-green to a rich yellow, often with red coloration. Areas adjacent to veins turn yellow last.

Uniform yellowing without green-veining.

Reddening, with areas near large veins remaining green longer.

Old E-4 leaf blade with yellowing, reddening and necrosis.

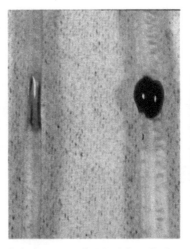

Colorless negative test for nitrate with diphenylamine reagent on cut petiole surface (left), light- to dark-blue positive test (right).

Characteristic red calyx (red-cap) of nitrogen-deficient fruit.

Smaller, normal-appearing white roots from hydroponically grown nitrogen-deficient Tioga plants (left), non-deficient (right).

Fruit from nitrogen-sufficient Tioga plant (top row); fruit from nitrogen-deficient plant (bottom row).

Symptoms of sulfur deficiency differ very slightly from those of nitrogen deficiency. Leaf blades uniformly turn from green to light-green and finally yellow for both deficiencies. In nitrogen deficiency the blades and leaf stalks of the older leaves develop a reddish cast, and the younger leaves actually become greener with increased nitrogen deficiency. In contrast, all leaves of sulfur-deficient plants tend to remain uniformly yellow until advanced stage of deficiency, when small, necrotic areas develop on the leaf blades.

Nitrogen-deficient plants can be readily distinguished from sulfur-deficient plants by testing the petioles for nitrate with diphenylamine reagent or by analyzing the leaf blades for sulfate-S. Sulfur-deficient blades will contain less than 100 ppm of sulfate-S (dry basis) as compared to more than 100 ppm for plants without a sulfur deficiency. Sulfur deficiency has no appreciable effect on the appearance of the fruit, except for reduced size.

Corrective measures

Incorporate gypsum at 560 to 1,120 kg/ha (500—1,000 lb/acre), either as a side dressing or within the bed before planting. This application of gypsum (Ca $SO_4 \cdot 2H_2O$) is equivalent to 65 to 130 g/m of row (65—130 oz/100 ft). For S, the rate becomes 12 to 24 g/m (12—24oz/100 ft). (See also page 55).

Uniform yellowing (light-green color) of leaves with moderate sulfur deficiency on Shasta plants.

Young to old leaves (left to right) showing uniform yellowing of sulfur-deficient (upper) and normal greening of sulfur-sufficient (lower) Shasta plants.

Uniform yellowing of sulfur-deficient, young-mature Shasta leaf.

Comparison of young-mature sulfur-deficient (left) and normal (right) leaflets of Tioga.

Close-up of uniform yellowing of sulfur-deficient Shasta leaflet.

Yellowing and drying with age of Tioga leaflets.

Sulfur deficiency (left) has no appreciable effect on root development (normal, at right).

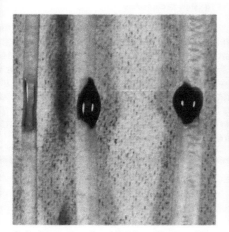

Colorless negative test for nitrate with diphenylamine reagent on cut surface of petiole from nitrogen-deficient plant (left), blue positive test for nitrate for petioles from sulfur-deficient (center) and normal (right) plants.

Normal fruit from sulfur-deficient and sulfur-sufficient plants. Calyx (cap) not shown, is also normal.

Early symptoms of deficiency are very similar to those of nitrogen, but even more of sulfur. However, unlike nitrogen symptoms, small, necrotic areas ultimately develop (as they do with sulfur deficiency) in the young mature leaf blades, followed by scorching and upward curling of the upper blade margins.

Chemical tests help to separate the causes of uniform leaf-blade yellowing. Petioles of both molybdenum- and sulfur-deficient plants test strongly positive for nitrate with diphenylamine reagent, thus eliminating nitrogen as the deficient element. Analysis of leaf blades for sulfate-S will indicate if yellowing is caused by sulfur deficiency. Likewise, a molybdenum leaf blade value of less than 0.4 ppm, dry-basis, or less, will indicate deficiency. Neither fruit size nor quality is affected appreciably by a mild deficiency of molybdenum.

Corrective measures

Plants suspected of being Mo-deficient should be treated on a trial basis with a foliage spray containing ammonium or sodium molybdate ($Na_2MoO_4 \cdot 2H_2O$) at 1.5 g/liter (1.25 lb/100 gal) of water, plus 0.1% detergent as spreader, or by the addition of either salt at the rate of 0.56 to 1.12 kg/ha (8—16 oz/acre) or 0.065 to 0.13 g/m (0.065—0.13 oz/100 ft) incorporated with fertilizer applied either at planting or later as a side dressing. (See also page 55).

Moderate deficiency of molybdenum with uniform yellowing of young-mature Tioga leaves and necrosis of older leaves. No characteristic symptoms appear on flowers, fruit, or roots.

Uniform yellowing, of youngest Tioga leaf (upper left), necrosis of mature leaves and no symptoms of an old leaf (lower right).

Middle leaflet, Tioga age series. Youngest leaflet (upper left), oldest leaflet (lower right).

Uniform yellowing of young-mature leaflets of Shasta (left), E-4 (center), and Tioga (right).

Initial collapse of mesophyll Tioga blade tissue.

Necrotic (dead) spots (drying) of mesophyll Tioga blade tissue.

Pronounced molybdenum deficiency symptoms on leaflet of Tioga.

Tioga leaflet with severe molybdenum deficiency symptoms.

Yellowing, collapse and drying of Tioga mesophyll blade tissue and start of marginal leaf scorch.

Deficiency symptoms are recognizable only by careful inspection of plants. They can be confirmed by laboratory analysis when acetic acid soluble phosphate-phosphorus values are less then 700 ppm, dry basis, for petioles of dark-green leaves. At the onset of phosphorus deficiency, plants have an overall deep-green appearance, with leaves somewhat smaller than normal. As the deficiency becomes more severe, upper surfaces of leaf blades develop a metallic lustre, almost a shoe-polish black for some varieties. Lower surfaces of leaf blades develop a reddish-purple cast, which may extend to the upper surfaces of leaf blades of some varieties, as leaves become older. Flowers and fruit of phosphorus-deficient plants tend to be smaller than normal, and fruit develops albinism occasionally in susceptible varieties. Root development appears normal initially but is not abundant, just as for nitrogen deficiency; also top growth is retarded much more than root development.

Corrective measures

Addition of P at 50 kg/ha (45 lb/acre) or 5.8 g/m (5.8 oz/100 ft) at planting or as a side dressing when the deficiency first appears. (See also page 55).

Stunted greening of phosphorus-deficient Tioga plant.

Dark-green color and black sheen on upper surface of phosphorus deficient Tioga leaf.

Comparison of upper surfaces of leaves from Shasta (upper row) and E-4 (lower row) from nitrogen-deficient (left), phosphorus-deficient (center), and normal (right) plants.

Characteristic purpling of lower surface of phosphorus-deficient Tioga leaf.

Upper surface of young-mature leaflets with characteristic dark greening of phosphorus deficiency (upper left) compared with the yellowing symptom of nitrogen deficiency (upper right). Old-mature Tioga leaflets redden with age from phosphorus deficiency (lower left) and nitrogen deficiency (lower right).

Characteristic purpling of lower surface of Tioga leaflet with phosphorus deficiency (left) and reddening with nitrogen deficiency (right).

Blue-green, blackened upper surface of phosphorus-deficient leaflet.

Stunted, darkened roots with phosphorus deficiency (left) compared to Tioga roots with ample phosphorus (right).

Fruit symptoms vary with variety: (E-4), with symptoms of albinism (left); Shasta, normal appearing; and Tioga with red calyx (cap).

21

Deficiency symptoms vary slightly with strawberry variety, climate, and soil, and are frequently confused with deficiencies of magnesium—or with leaf scorch caused by sodium chloride, sun, wind, drought, disease, or pest damage. The first symptom of potassium deficiency often appears as a tanning, browning, and drying process of the upper margin of a young, mature leaf blade. This injury progresses inwardly between the veins to include most of the leaf blade. Almost simultaneously, the lower blade area, including the mid-rib and short petiole section from the blade to the leaf stalk, darkens and becomes dry. This symptom is unique to the strawberry plant and identifies the problem as a shortage of potassium. Although older leaves may be severely affected, younger leaves remain free of symptoms. Apparently, potassium is translocated from older to younger leaves, and at times the newly translocated potassium may be sufficient for a substantial amount of new leaf growth. Bright, sunny weather exacerbates leaf scorch, thus the symptom is often erroneously associated with sunburn. Petioles of scorched blades develop light- to dark-brown longitudinal lesions, then they wither and collapse. Fruit fails to develop full color, is pulpy in texture, and insipid in taste. Fibrous roots darken in potassium-deficient nutrient solutions but recover upon addition of potassium.

Sodium as an ion is not ordinarily absorbed by the strawberry plant. When it is, it is a sure sign of root injury. This failure to absorb sodium also results in retention of potassium by the petioles even when blades are severely deficient in potassium. Consequently, if potassium deficiency is suspected, blades should be analyzed for potassium. A leaf blade with potassium content of less than 0.5 per cent indicates deficiency.

Corrective measures

Treat with potassium sulfate at 95 kg/ha (85 lb/acre) or 11 g/m of row (11 oz/100 ft) at planting or as a side dressing. (See also page 55).

Potassium-deficient. Symptoms begin on young-mature Shasta leaves (leaf to the right) with a tanning marginal scorch, followed by a darkening and necrosis of the lower mid-rib area, a unique symptom observed thus far only in the strawberry. Petioles are also brown.

Potassium deficiency often begins with a tanning, darkening, and marginal scorch of the blade tissue, followed by a darkening of lower mid-rib area.

Symptoms on Tioga increase in severity with leaf age (upper left to lower right), proceeding from a nearly normal green to a marginal scorch and nearly full scorch (lower right).

Tanning, marginal scorch and browning on Tioga of the lower mid-rib area of potassium-deficient leaflet.

Close-up of tanning and basal mid-rib browning on E-4 of potassium-deficient leaflet.

Close-up of tanning and browning of potassium-deficient Tioga leaflet.

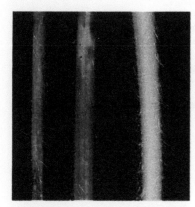

Light-green color (right); freckling and browning (center); and drying (left) of potassium-deficient petioles.

Potassium-deficient Tioga roots appear normal, except for a dark pigmentation (left) which fails to appear with sufficient potassium (right).

Potassium-deficient Tioga fruit is light in color, soft, and tasteless.

Scorching of mature blades from magnesium deficiency begins as yellowing and browning of the upper margins of blade tissues. Progressing inwardly, the interveinal areas become chlorotic, then necrotic, to form a blotchy leaf pattern. This is followed by further scorching, with the basal portion of the blade remaining light-green and turgid to the last. With potassium deficiency, on the other hand, the basal portion scorches first. The young, central leaves remain green, just as with potassium deficiency. The short petiole sections between blades and leaf stalk, instead of darkening and drying, remain turgid and green. In both deficiencies, scorching increases with increased deficiency and leaf age. Deficient blades contain less than 0.1 percent magnesium, dry basis. Fruit from magnesium-deficient plants appears nearly normal, except for a lighter red color and a tendency to albinism. Magnesium-deficient roots appear normal, except for an overall reduction in quantity.

Corrective measures

Fertilization with materials containing readily available magnesium, such as magnesium sulfate applied at planting or later as a side dressing at 56 to 112 kg/ha (50—100 lb/acre) or 6.5 to 13.0 g/m (6.5—13.0 oz/100 ft of row), followed by an increase in magnesium uptake, will prevent leaf scorch. (See also page 55).

Magnesium deficiency on E-4 first appears as a leaf scorch on young-mature leaves.

Marginal scorch (left), and normal young-mature leaf (right) of Tioga.

Severe scorch on old-mature Shasta leaf.

Youngest Tioga leaflets are often without symptoms (upper left); severe leaf scorch occurs on oldest leaflet (lower right).

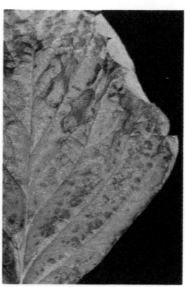

Browning, followed by collapse of marginal and sub-marginal Tioga tissues.

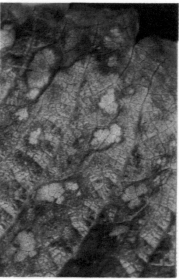

Close-up of marginal and sub-marginal scorch on Tioga leaflet.

Magnesium-deficient Tioga roots (left) and normal roots (right).

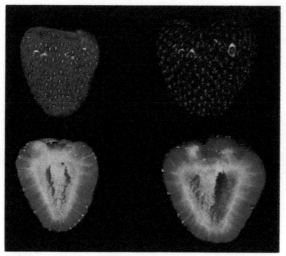

Magnesium-deficient Tioga fruit (left) is light-red and is soft in texture compared to normal fruit (right).

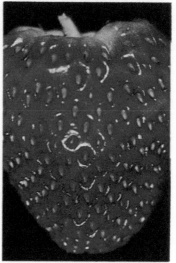

Close-up of light-colored magnesium-deficient Tioga fruit.

Cl and Na

Salt damage on old Aiko leaves from only 0.5 me per liter of sodium sulfate.

Concentric marginal scorch rings on Tioga from sodium chloride.

The need for chlorine or for sodium has not been established for the strawberry plant. Even plants grown in virtually chlorine-free nutrient solutions and in chlorine-free, carbon-filtered air, show no deficiency symptoms. This implies that there is very little if any chlorine requirement for strawberry. Similarly, sodium is not required for the growth of the strawberry.

Because chlorine and sodium are not readily absorbed from nutrient solutions or from soils by the strawberry plant, it is often referred to as a chloride-and-sodium-excluder. This exclusion may account for the apparent non-essentiality of these ions for the strawberry plant and for some protection against salt damage in its seashore habitat. Unfortunately, protection is only partial, since plants are readily subject to "salt damage" when grown in soils relatively high in salinity. "Salt damage" is apparently due primarily to sodium, which is increased by chloride when present in equivalent amounts and is diminished by soluble calcium when it is added as gypsum to the soil. Leaf blades with "salt burn" symptoms and with concentrations greater than 0.10 percent sodium or 0.50 percent chlorine, indicate salt damage.

Corrective measures

A review of local water quality, soil salinity, and drainage problems should be made in cooperation with technical advisors before corrective measures for salt damage are undertaken by the grower.

"Salt" damage on Aiko from NaCl (left) and Na₂SO₄ (center) greatly exceeds that from CaCl₂ (right). Sodium chloride at a concentration of 32 milliequivalents (me) per liter in half-strength complete nutrient solution damages plant growth greatly (left). Growth is not improved appreciably by the replacement of chloride by sulfate (center). It is, however, greatly improved by the replacement of sodium by calcium (right).

Chloride has little effect on plant growth of Aiko. Compare calcium chloride at 0.5 me per liter (left) to 32 me per liter (right).

Salt burn on mature Aiko leaves from sodium chloride (left), sodium sulfate (center) and from calcium chloride (right) at 32 me per liter of nutrient solution.

Root damage on Aiko from sodium, as sodium sulfate (32 me) (left) or sodium chloride (not shown), greatly exceeds that from equal chloride added as calcium chloride (right).

Normal Aiko leaflet (left) compared to those with salt burn from sodium chloride, sodium sulfate or calcium chloride (left to right).

Marginal scorch on Aiko may occur with sodium sulfate as low as 1 me per liter.

Deficiency of calcium induces a greater number of characteristic symptoms than does any other nutrient element. It is responsible for "tip-burn," hardening of fruit, root-tip stunting, and growing-point damage. Tip-burn appears most frequently during periods of rapid leaf growth and occurs in some varieties much more frequently than in others. Blades on plants becoming calcium-deficient, are crinkled, puckered, and have light-green or light-yellow borders. The next blades to form are also crinkly, and the tips fail to expand fully, becoming black, to form the characteristic symptom known as tip-burn. Leaf stalks of these leaves freckle, frequently exude globules of syrup, and collapse near their mid-points. Similar symptoms may also appear on flower stalks, about a third of the distance down. Calcium-deficient fruit develops a dense cover of achenes either in patches or over the entire fruit; it is hard in texture and acid to the taste. Roots, instead of developing normally, become short and stubby and darken with age.

Leaf symptoms may also appear in the nearly fully developed blades of older leaves. These symptoms develop as light-green to yellow areas that coalesce in time and later become dry. During this process, globules of syrupy liquid frequently form on the mid-ribs of the blades.

Blades with characteristic tip-burn symptoms frequently contain less than 0.2 percent calcium, dry basis.

Corrective measures

Trial additions of gypsum ($CaSO_4 \cdot 2H_2O$) at 560 to 1,120 kg/ha (500—1,000 lb/acre) or 65 to 130 g/m of row (65—130 oz/100 ft) before planting or as a side dressing should be helpful, although most plants recover as growth rate and demand decrease during summer. (See also page 55).

To convert gypsum values to those of Ca, multiply by 0.232, e.g., 560 × 0.232 becomes 130 kg of Ca.

Calcium deficiency damages the growing point, induces tip-burn and causes chlorosis of Tioga leaves.

Browning of developing center leaf leading to tip-burn damage. This symptom frequently appears during hot weather.

Mature leaf with black-tipped, crinkled leaflets characteristic of tip-burn symptoms seen in the field.

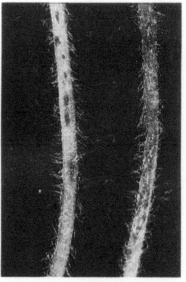

Mild tip-burn on Tioga. Leaflet with marginal chlorosis and sub-marginal crinkling (right). Pronounced tip-burn. Leaflet with black-tip and pronounced crinkling (left).

Syrupy globs on under surface of dark-green blade-tissue of calcium-deficient E-4 leaflet.

Brown freckling and longitudinal lesions on petioles of older Shasta leaves.

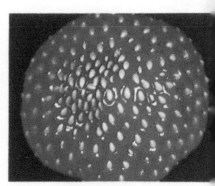

Short, stubby Shasta roots (left) and normal, filamentous roots (right).

Dense cover of seeds (achenes) on unexpanded fruit (left), compared to normal Tioga fruit (right).

Densely seeded area near fruit tip that could easily be confused with lygus bug injury on Tioga.

Boron and calcium deficiency symptoms are similar during the early stages of development. Both affect younger leaves, produce puckering and tip-burn, deform fruit, and dramatically shorten rootlet growth. Rootlet growth becomes stunted as soon as boron becomes depleted; shortly thereafter, marginal yellowing and crinkling of the younger leaves occur, and growing point growth is reduced. A moderate deficiency of boron reduces flower size and decreases pollen production, resulting in small, "bumpy" fruit of poor quality. This symptom differs from calcium deficiency, which decreases the spacing between achenes and excessively increases fruit firmness. In moderate boron deficiency, interveinal areas of leaf blades become chlorotic, whereas they remain green with calcium deficiency. Upward cupping of the leaf blade is less pronounced with boron than with calcium deficiency. Boron-deficient leaf blades contain less than 25 ppm of boron, dry basis.

Corrective measures

Make small-scale trials only, because strawberries are sensitive to slight excesses of boron applied either as a foliage spray or soil treatment. Beneficial and detrimental effects with boron will vary with variety, stage of growth, climate, and soil conditions. Unless local experience dictates otherwise, spray the foliage with B at the rate of 0.113 to 0.339 g/liter of water (1.5—4.5 oz/100 gal) or as a soil treatment at the rate of 0.66 g/m of row (0.66 oz/100 ft) applied at planting or as a side dressing when the symptoms first appear. During bloom, reduce B content of spray to 10 percent of that given for foliage (See also page 55).

Boron and calcium deficiency are similar; both damage the growing point, cause tip-burn, and form stubby rootlets but differ in kind of chlorosis and fruit damage on Tioga.

Tip-burn of youngest Tioga leaf (upper right) and oldest leaf (right) showing interveinal chlorosis.

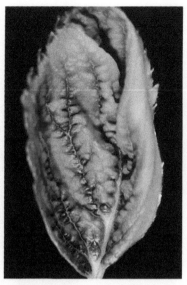

Close-up of tip-burn and interveinal chlorosis of boron-deficient Tioga leaflet.

Close-up of interveinal chlorosis on Tioga.

Boron deficiency causes bumpy E-4 fruit (due to poor pollination) and a raised, fleshy white collar.

A flower from a normal Tioga plant (left), boron-deficient (right).

Close-up of boron-deficient Tioga flower same as in photo to the left.

A severely boron-deficient fruit cluster of E-4.

Short, stubby, dark Tioga roots are similar to calcium deficiency (left) and normal filamentous roots (right).

Yellowing or chlorosis of young leaf blades is the first, although uncertain, sign of iron deficiency in the strawberry plant. As the deficiency becomes more intense, yellowing increases to the point of bleaching. Splotchy browning of the bleached blade tissues then will occur if deficiency is acute.

When deficiency is mild, veins—including veinlets—turn green. This "recovery" is frequently regarded as characteristic of iron deficiency. Chemical analysis for ion at this time may be misleading, because of recent uptake of iron. Also, when the chlorosis is due to zinc or manganese deficiency, iron accumulates in the blades. Surface dusts containing iron may also give high iron values for iron-deficient leaves. Iron deficiency has little effect on fruit, even when the deficiency is severe.

Less than 40 ppm iron (dry basis) for leaf blades indicates iron deficiency. Foliar sprays with non-toxic iron containing compounds or appropriate soil amendments with ferrous sulfate or chelated iron will ordinarily increase the iron content of leaf blades and correct the deficiency.

Corrective measures

Spray treatments with Fe at 0.1 to 0.2 g/liter of water (1.5 to 3.0 oz/100 gal) on foliage should be made on a trial basis only and with caution, because damage to flowers or fruit may possibly occur. Soil treatments with Fe may also be tried by applying iron sulfate (Fe $SO_4 \cdot 7H_2O$) or chelated iron at 1 to 2 g/m (1—2 oz/100 ft) of row applied at planting or as a side dressing when the symptoms first appear. (See also page 55).

Yellowing and green-veining of young, mature Tioga leaves and normal flowers are characteristic of iron deficiency.

Iron deficiency appears first as a yellowing (chlorosis) of young Tioga leaves, (first leaflet, left) and as a green-veining on young-mature leaves (second and third leaflet) or as a pale-green color (fourth leaflet).

Bleaching and marginal necrosis of young-mature leaflet occurs with extreme iron-deficiency of Tioga.

Yellowing and green-veining of young-mature Tioga leaf.

Yellowing, green-veining, with marginal and interveinal browning (necrosis) of Tioga leaflet.

Close-up of yellowing, green-veining and netted-veining, with marginal and interveinal necrosis of Tioga.

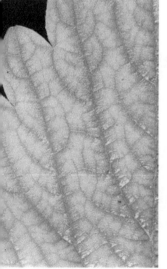

Close-up of yellowing, green-veining and netted veining of young-mature Tioga leaflet.

Reduced growth and accumulation of yellow-colored riboflavin by roots of deficient Tioga plant (left) compared to normal (right).

Fruit from iron-deficient Tioga plants (lower row) differs only slightly in size and number of fruit produced from normal plants (upper row).

Zinc deficiency can easily be detected by the green "halo" that appears along the serrated margins of young, immature leaf blades. This "halo" disappears with mild deficiency but persists with leaf age, when deficiency is pronounced. As leaves continue to grow, blades become narrow, especially at the base, and become quite elongated with severe deficiency. As a rule, necrosis does not occur even with severe deficiency. As the leaves age, veins and surface tissues redden—a pronounced characteristic in some varieties. Zinc deficiency tends to increase fibrous root length at the expense of total root growth. Fruit development appears normal, although number and size of fruit are reduced. Deficient blades contain less than 10 ppm of zinc, dry basis.

Corrective measures

Try a spray treatment with materials containing Zn at 0.2 to 0.4 g/liter of water (2.5—5 oz/100 gal). Spray may damage young strawberry foliage, flowers, or possibly fruit, for this reason, test only a few plants or plots. Soil treatments are safer and last longer, but recovery of plants will be slow to start. As a trial, treat the soil with Zn from zinc sulfate ($ZnSO_4$) or chelated Zn at the rate of 1 to 3 g/m of row (1—3 oz/100 ft), applied at planting or as a side dressing, followed by irrigation, when the symptoms first appear. (See also page 55).

Severely zinc-deficient Shasta plant with young leaflets showing yellowing, green-veining, reddening, and a serrated green halo.

Young, yellow E-4 leaves with green-veining, mild necrosis and marginal greening. Marginal greening in the form of a serrated green "halo," is unique to zinc-deficient strawberries.

Basal narrowing of leaflets also occurs with the yellowing, green-veining, and green "halo" symptoms on Tioga.

Reddening may occur from zinc deficiency in some varieties.

Leaflet with yellowing, reddening, green-veining, and serrated green "halo" on Shasta.

Close-up of yellowing, green-veining, serrated green "halo" and base of Tioga leaflet narrowing.

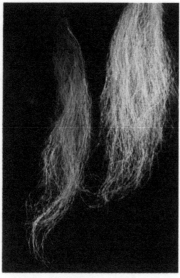

Close-up of zinc-deficient leaflet with reddening, yellowing, and green-veining.

Filamentous Tioga roots of zinc-deficient plant (left), normal (right).

Nearly normal Tioga fruit from zinc-deficient plant (left), and normal fruit (right).

The first sign of deficiency is yellowing of young, developing leaves. Symptoms are similar to the overall, light-greening with molybdenum, sulfur, and iron deficiencies. In time, the light-greening develops into very fine, netted veining or clear dotting, that is unique to manganese deficiency. As deficiency increases, the main veins remain dark-green, with interveinal areas becoming light-green to yellow, followed by scorching and upward turning of the leaf blade margins. Scorching moves progressively inward as a series of broad rays, extending across veins, in contrast to the interveinal scorching of iron deficiency. Except for a decrease in fruit size, manganese deficiency has no appreciable effect on fruit appearance or quality. Leaf blades with less than 25 ppm manganese, dry basis, indicate deficiency.

Corrective measures

Conduct trials with a foliage spray containing Mn at 0.08 to 0.16 g/liter of water (1.0 to 2.0 oz/100 gal) or as a soil treatment at the rate of 1 to 2 g/m of row (1—2 oz/100 ft) applied as manganese sulfate or manganese chelate as a side dressing when the symptoms first appear or at the next planting. Do not spray at bloom time or at heavy fruit set, except for small-scale trials. (See also page 55).

Manganese-deficient Shasta with yellowing of young mature leaves and severe marginal and interveinal necrosis of older leaves.

Pale-greening to yellowing of young-mature manganese-deficient Shasta leaf.

Severe marginal and interveinal necrosis of manganese-deficient Shasta leaf.

Netted, clear-dotted veining symptom on E-4, unique to manganese deficiency.

Close-up of netted-veining on Tioga, unique to manganese deficiency.

Close-up comparison of manganese (left) and zinc deficient leaflets (right).

Close-up of older-mature, manganese-deficient Shasta leaflet.

Shasta fruit appears normal, except for smaller size, even when plants are severely deficient in manganese.

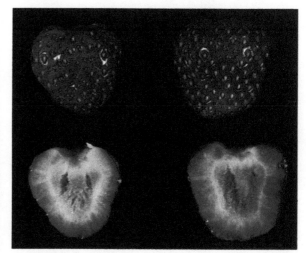

Manganese deficiency (left) reduces yield and at times fruit size, but has no appreciable effect on Tioga fruit appearance.

37

The first symptom of deficiency is a uniform light-green color of young, immature leaf blades, similar to the early symptom of sulfur, molybdenum, manganese, and iron deficiencies. Shortly thereafter, interveinal areas become very light-green with primary veins remaining initially conspicuously green. Gradually, interveinal areas and veins, except for a broad, green border, become bleached. Bleaching, without green veining, is also characteristic of copper deficiency in sugarbeet and cereal crops. No specific symptoms of copper deficiency were observed on roots or fruit of strawberry. Leaf blades with less than 3 ppm of copper, dry basis, indicate copper deficiency.

Corrective measures

Try a Cu material (copper sulfate or copper chelate) applied to small areas either as a foliar spray (0.075—0.15 g/liter, 1—2 oz/100 gal water) or as a soil amendment (1—2 g/m of row; 1—2 oz/100 ft of row). (See also page 55).

Copper-deficiency first appears as a uniform light-green color, without a serrated green ring on the blades of young Shasta leaves.

Light-green young leaflets (left); bleaching, green-veining, and green border of mature leaflet (center); and dark-green old Shasta leaflet (right).

Light-green, young Shasta leaflet, without a green border but with some green-veining.

Mature Shasta leaflet with bleaching, green-veining, and pronounced green border.

Close-up of bleached area, green-veining and green border on Shasta.

Close-up of bleached area, a symptom unique to copper deficiency of plants.

Shasta fruit without symptoms.

Sometimes strawberries produce white berries that are normal in size and general appearance but later fail to color properly, are insipid, mushy, unattractive, and spoil rapidly after picking. Externally, the skin shows white areas, is tender, and bruises easily. Internally, the fruit is mottled, pink and white. "Off" color and poor shipping qualities of such fruit result in serious losses of saleable berries and are disappointing to the home gardener and consumer. A shortage of sugar during fruit ripening is the primary cause of the white berry (or albino fruit disease) of strawberries.

Corrective measures

Introduce varieties more adapted to local climatic conditions; disease and pest control might be improved; nitrogen should be applied to just meet the needs of the plants, and the flow of sugar to the fruit may be stimulated by hormonal sprays. (See also Albinism, page 52.)

Albino Tioga fruit produced by plants with small tops induced by inadequate winter chilling. (South coastal inland).

Albino fruit produced by plants with large tops during a cloudy, low light period compared to normal fruit (lower left corner). (Watsonville).

Albino fruit produced in plant growth chamber at low-light intensity.

Albino fruit produced at a low-light intensity of 500 foot candles.

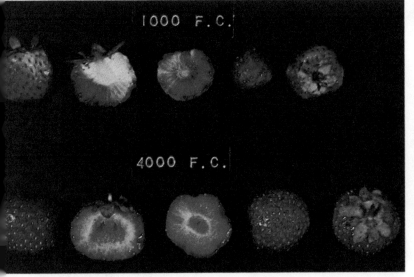

Albino Shasta fruit produced at 1,000 foot candles starting at flowering (top row) compared to normal fruit at 4,000 candles (bottom row) for 14-hour photoperiods.

Close-up of albino fruit compared to normal fruit.

Albino Shasta fruit variation at 1,000 foot candles.

Seeds of albino E-4 fruit are surrounded by red and then white flesh.

PLANT ANALYSIS APPROACHES TO DETERMINE FERTILIZER NEEDS

A strawberry grower may wait until deficiency symptoms appear on his plants before fertilizing or he may choose to prevent the appearance of symptoms by fertilizations based on information gained during the growing season from the use of the systematic plant analysis program discussed later in this chapter.

Confirmation of visual diagnosis by chemical analysis

If, by chance, symptoms have already appeared, one may first compare them to those depicted in the Color Atlas to tentatively determine their cause; next, the visual diagnosis should be confirmed by comparing the analytical values for the affected leaves to those taken from plants with known deficiency symptoms in Table 1. Finally, the deficiency—or deficiencies—can be corrected with appropriate fertilization.

If, for example, the leaf symptoms appear to be due to a deficiency of nitrogen, the question "Are my strawberries getting enough nitrogen?" can be answered almost immediately by using the diphenylamine test for nitrate on petioles of newly matured leaves as described in the next section. For a precise evaluation, the petioles may be chemically analyzed in the laboratory for nitrate-N. For other deficiencies the blades are preferable for analyses.

Diphenylamine test for nitrate. In this test, a drop of diphenylamine reagent is placed on the cut surface of a slantwise cut through the petiole of a leaf (Color Atlas, page 15). If a distinct blue color develops immediately, the nitrate-N values are more than 500 ppm, and the plant is adequately supplied with nitrogen at the time of the test. If the drop of reagent remains colorless or turns blue very slowly, the nitrate-N values are less than 500 ppm, and the plant is nitrogen-deficient at the time of sampling.

TABLE 1. Plant Analysis Values for Determining the Mineral Status of Strawberries (Dry Weight Basis)*

Nutrient	Plant part tested	Tentative critical concentration	Range showing	
			Deficiency symptoms	No deficiency symptoms
Boron: B	Blade	25 ppm	18-22	35—200
Calcium: Ca	Blade	0.3%	0.08-0.20	0.4—2.7
Chlorine: Cl	Petiole	---%	<0.07	0.07—0.4†
Copper: Cu	Blade	3.0 ppm	<3.0	3—30
Iron: Fe	Blade	50 ppm	5-40	50—3000
Magnesium: Mg	Blade	0.2%	0.03-0.10	0.3—0.7
Manganese: Mn	Blade	30 ppm	4-25	30—700
Molybdenum: Mo	Blade	0.5 ppm	0.12-0.40	0.5+
Nitrogen: NO_3^-–N	Petiole	500 ppm	0-500	700—20,000
Total —N	Blade	2.8%	2.0-2.8	3.0+
Phosphorus: $H_2PO_4^-$–P	Petiole	700 ppm	150-700	1,000—5,000
Total —P	Blade	1,000 ppm	300-1100	1,500—13,000
Potassium: K	Petiole	1.0%	0.1-0.4	1.0—6.0
	Blade	1.0%	0.1-0.5	1.0—6.0
Sodium: Na	Blade	---%	<0.01	0.01—0.4†
Sulfur: SO_4–S	Blade	100 ppm	25-80	100—500+
Total —S	Blade	1,000 ppm	300-900	1,000+
Zinc: Zn	Blade	20 ppm	6-10	20—50+

*NOTE. Before analyzing leaf blades for microelements, dust, soil, fertilizer, and spray residues must be removed. This can be done by washing the leaves for 30 seconds in a bath containing a detergent or in one containing a detergent and 0.1 N HCl, followed by two successive rinses in distilled water. The nondeficient range of values was determined on the young, fully expanded leaves. The leaves in the deficiency range varied in age, but were always taken shortly after the symptoms appeared.

The upper value reported is the highest value observed to date for normal tissues. Abnormally high values are often associated with other nutrient deficiencies; for example, blades low in Zn may be high in Mn or Fe, and so forth.

†Salt damage is indicated by values above 0.50 percent for Cl and/or 0.10 for Na.

Diphenylamine reagent is prepared as follows: Add 0.2 gram of diphenylamine to 100 ml of nitrate-free, concentrated sulfuric acid. Store reagent in a fully-labelled pyrex glass-stoppered bottle and keep in the dark. Dispense small amounts as needed from a dark-colored glass-stoppered eyedropper bottle for field use.

CAUTION: Sulfuric acid is highly caustic. Keep away from children and animals. If reagent contacts skin or clothing, wash immediately with large amounts of water. Follow this with a dilute solution of baking soda. Always pour acid or reagent into water **not** the reverse.

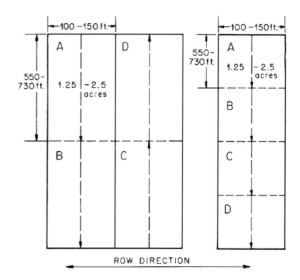

Fig 2. Sampling procedure. Divide a 5- to 10-acre block into four sections, A, B, C, and D. Walk down the center of section A, taking a recently matured leaf from a plant at equal row intervals, for a total of 30 to 40 leaves per section. Repeat for sections B, C, and D. For further instructions, see text.

Comparison of leaves with and without symptoms.

Questions about the adequacy of other nutrients—phosphorus, potassium, magnesium, calcium, iron, zinc, manganese, copper, and boron—can also be answered by chemical analysis of comparable leaf blades, with and without symptoms (Table 1). The methods used for these nutrients are far more sophisticated than for nitrate, but in a modern, well-equipped laboratory they can be determined easily with only 100 mg samples of dry, ground blade material. Molybdenum determination, however, requires from 1 to 5 grams of material whereas a sample of only 25 or 50 mg is required to test for sulfate-sulfur deficiency.

Systematic plant analysis program to prevent deficiencies

Unfortunately, by the time nutrient deficiency symptoms have appeared in the field, economic losses have quite likely occurred. To prevent such deficiencies and to maintain desired nutrient levels, a systematic plant analysis program is recommended to monitor the plants under field conditions. The current system of monitoring strawberries for nitrogen and other essential minerals has been developed by the authors from experience gained primarily with sugarbeets, grapes, and tomatoes, and, more recently, with 'Shasta,' 'Tioga,' and 'Tufts' strawberry varieties at the Deciduous Fruit Field Station in San Jose, California. The system consists of taking petiole and blade samples (Figures 2 and 3) about every 2 weeks during the growing season for chemical analysis. By plotting the analytical results vertically on a graph against time horizontally, changes in the mineral concentration of the leaves can be compared to the critical concentration and nutrient level desired for that time of the growing season (Figure 4). In This way the strawberry grower

can keep track of the nutrient status of his crop and, if necessary, make adjustments in his fertilizer program, up or down, based on the needs of the crop, either currently or for succeeding crops. For example, for best growth, it may be essential to apply relatively small amounts of nitrogen frequently; or perhaps nitrogen should be released slowly, from organic or encapsulated materials, or continuously by drip irrigation according to the changing needs of the crop. Whatever fertilization is required, the monitoring costs, especially for nitrogen, will be offset under most conditions by increased production of quality fruit, more effective use of nitrogen, and movement of less nitrate into the ground water.

Sampling procedure. The petiole of a young mature leaf (Figure 3) is selected for nitrate, chloride, and 2 percent acetic acid-soluble phosphorus analysis, and the bladelets of this leaf are used for all other analyses—potassium, calcium, magnesium, iron, manganese, zinc, copper, molybdenum, boron, total nitrogen, total phosphorus, and sodium.

Figure 3 shows a diagram of a sampling unit and gives instructions for collecting leaves. A composite plant sample for chemical analysis usually consists of 30 to 40 leaves, each taken at equidistant intervals across the rows of a sampling unit. Such a composite plant sample is usually taken across the rows in the middle of one-quarter of a square-

Fig. 3. Selection of leaf to sample for chemical analysis. **A.** View of leaves from the top of the plant. **B.** Side view of plant showing immature, young-mature, and old leaves. (Arrows indicate young-mature leaves to sample.) **C.** Age array of leaves. First three leaves, left to right (top row) are immature; fourth and fifth leaves are young-mature. Bottom row, old leaves.

For analysis, separate the three leaflets from the leaf stalk of young-mature leaf sample, and place them into pre-labelled paper (not plastic) bag; do likewise for the leaf stalk. Keep samples cool prior to drying at 70° to 80°C in a forced draft oven.

shaped field, or across the middle of each quarter of a long, narrow field. Four samples are taken this way from a 5- to 10-acre field at intervals of 1 to 3 weeks for the first year and thereafter as required to evaluate the fertilizer program adequately.

Leaves, as they are collected in the field, are separated into petioles and blades (leaflets), placed in small pre-labelled paper bags, and kept cool until placed in a vented oven to dry overnight at about 700C. Dusty leaflets to be analyzed for iron, manganese, zinc, copper, or molybdenum must be washed in a weak acid solution containing a detergent and rinsed in distilled water before drying. After drying, the samples are ground in a Wiley mill, or one similar to it, so as to pass a 20- or 40-mesh screen, transferred to a labelled plastic vial, and capped to await chemical analysis.

Analytical methods. Labor intensive analytical methods involving wet chemistry and low cost equipment have been superseded by precise, rapid methods, which however, utilize costly equipment, such as the AA (Atomic Absorption) analyzer for the determination of K, Na, Ca, Mg, Fe, Mn, and Zn, and the ICP (Inductively Coupled Plasma) spectrometer to determine these same cations plus Cu,

total-P, total-S and total-B. The anions NO_3-N, SO_4-S, H_2PO_4-P and Cl can now be determined easily by ion chromatography (IC). Total-N, which in the Kjeldahl procedure involves the determination of ammonium, can now be far more quickly determined directly in the digestate without distillation by an auto-analyzer. Each instrument requires only 50 or 100 mg of finely ground plant material, passing a 30 or 40 mesh sieve, for analysis. With a sand or metal-block wet digester, 32 to 96 samples can be analyzed routinely by AA or ICP with computer printouts. The wet digestion requires a pretreatment, usually overnight, with concentrated nitric acid in graduated Taler digestion tubes, followed either by nitric or nitric-perchloric acid (Johnson and Ulrich, 1959) to complete the digestion.

The diphenylamine test for nitrate-N, as described on pages 15 and 42, is, fortunately, a quick and easy way to determine, under field conditions, whether nitrogen, the key nutrient, is adequate for maximum fruit production when sampled periodically during the growing season. Nitrate-N may also be determined routinely in the laboratory by precise methods when 1 to 2 grams of dry petiole material are available for analysis.

TABLE 2. Effect of Nitrogen Fertilization on Petiole Nitrate-N and Yield of Summer-Planted Shasta Strawberries 1969-70.

Sample & Harvest Dates[a]	Accumulated N lb / A[b]		Petiole N NO₃-N ppm		Yield[c] tons / A		Change in Yield tons / A	% change
	without fertilizer	with fertilizer	without fertilizer	with fertilizer	without fertilizer	with fertilizer		
February 10	0	160	20	30	—	—	—	—
March 12	0	200	400	2260	—	—	—	—
March 24	0	—	730	2050	—	—	—	—
March 31	0	—	—	—	0.35[d]	0.18	−0.17	−48.6
April 7, 14	0	—	1460	4680	2.92	3.29	0.37	12.8
April 21, 28	0	—	480	3170	3.75	4.74	0.99	26.4
May 5, 12	0	240	10	1670	2.85	3.34	0.49	17.2
May 19, 26	0	—	0	2470	2.34	3.36	1.02	43.6
June 2, 9	0	—	0	1230	2.13	3.25	1.12	52.6
June 16, 23	0	—	250	370	3.08	5.07	1.99	64.6
June, July 29, 6	0	—	0	120	2.32	4.03	1.71	73.7
July 13, 20	0	280	0	0	1.72	2.26	0.54	31.4
August 3	0	—	0	130	—	—	—	—
Total Edible Fruit					21.44	29.51	8.07	37.6

[a] Dates when petioles were sampled and fruit harvested on the same date (Bold type).
[b] Ammonium nitrate was side-dressed at the rate of 40 lb/acre on 9/23, 10/21, 11/17/69, 2/10/70, accumulating to 160 lb/acre of N on 2/10/70, to 200 lbs. on 3/12, 240 lbs. on 5/5 and to 280 lbs. on 7/13/70 (See Fig. 4).
[c] Yields are based on four replications per treatment and are averaged for two picking dates except for the single picking on 3/31/70.
[d] Fruit from unfertilized plants ripens early.

NITROGEN: THE MOST FREQUENTLY NEEDED NUTRIENT

Nitrogen turned out to be the key nutrient in ongoing fertilizer studies with strawberries at the Deciduous Fruit Field Station in San Jose, California, that began in 1968. Without nitrogen fertilizer our plants ran out of nitrogen rapidly, and production decreased dramatically (Figures 4 and 5 and Table 2). With an ample supply of nitrogen, plant analysis showed our plants contained an adequate supply of all other nutrients—except possibly zinc, phosphorus, and calcium. Zinc and phosphorus were found to be low at times early in the growing season, probably because of low soil temperatures; calcium, judging by the tip-burn symptoms of leaf blades and hard-nosed fruit, appeared to be low only under high-nitrogen conditions, with some varieties affected more than others.

Our studies also showed that the systematic use of plant analysis is a useful tool in measuring the nutritional status of strawberry plants at all stages of development, including just after planting, overwintering, spring growth, blooming, fruit enlargement, ripening and dormancy after fruit picking. Nitrogen at all these times played a key role in fruit production. Whenever petioles fell below 500 ppm of nitrate-N, the plants were deficient in nitrogen whereas above 500 ppm they were well supplied with nitrogen at the time of sampling. Thus, by taking petiole samples any time during the growing season, the nitrogen status of the plants could be determined and fertilizer nitrogen applied as needed.

The relationship of petiole nitrate-N to yield and the effects of various cultural practices on petiole nitrate-N (and again on yield) are reported in the following sections of this chapter. Similar chapters could have been written about phosphorus, potassium, zinc, or any other nutrient, if any of these had become deficient after the plants' nitrogen needs had been met.

How to interpret nitrate-N values. By plotting the values for petiole nitrate-N, as done for an experiment with summer-planted strawberries in Figure 4, the nitrogen status of a strawberry plant can be visualized as a series of nitrogen deposits and withdrawals in a nitrogen back account. In the experiment, the threat of nitrogen bankruptcy for the strawberry plant started at the critical level of 500 ppm nitrate-N (dry basis) for petioles collected from young, mature leaves. Below that point, rates of growth and fruit production started to decrease (Figure 5).

It can be concluded that the earlier in the growing season a nitrogen deficiency occurs, and the longer it lasts, the greater is the loss of harvestable fruit. Conversely, the longer petiole-N values remain above the critical level, yet are not excessive, the larger the amount of harvestable fruit. However, increases in nitrate-N above 2,000 ppm do not result in more fruit (see Figures 4 and 6). Moderately high values of 3,000 to 10,000 can be viewed as nitrate reserves to be drawn upon later or as "safe values" for best fruit production. When nitrate-N values become very high (above 10,000 ppm), leaf growth becomes excessive, and yields are reduced. In actual practice, plants are usually in the 5,000- to 10,000-ppm range during rapid, vegetative growth, blossoming, and fruit set, followed thereafter by lower values.

While the critical value for nitrogen has been set at 500 ppm nitrate-N (dry basis) for petiole samples taken at any time of the year from young, mature leaves for all strawberry varieties, it is important to note that the critical value of 500 ppm is not the desired value to be achieved, but the one to remain well above, particularly during rapid vegetative growth after planting and during the period of flowering and fruit production. At these times, demand for nitrogen is high and petiole values for nitrate-nitrogen should be in the thousands and not even close to the critical value of 500 ppm—a point where rates of growth and fruit production begin to decrease.

Obviously, when petiole values fall below 500 ppm, nitrogen must be applied immediately to avoid fruit losses (see Figure 5). Also, in subsequent crops on the same or similar fields, larger amounts of readily available nitrogen should be applied earlier, more frequently, better placed, or the like, for better fruit production.

Constancy of the critical value. Since 1968, a concentration of 500 ppm nitrate-N, dry basis, has been used as the critical point to indicate nitrogen deficiency of strawberries under field conditions. This value has been found to be relatively constant, regardless of variety, year, time of sampling, plant spacing, or sampler. What changes, however, is the nitrate-N concentration of the petioles, which reflects immediately the nitrate-

supplying power of the soil and the nitrate demanded by the plant. The nitrate-supplying power of the soil varies with the composition of the organic matter added as manure or present from previous crop residues. These materials decompose and produce nitrate at various rates depending on soil temperature, moisture, and aeration (compaction). In the plant, the demand for nitrate also varies as growth rates change due to plant age, yearly differences in climate, or to differences in cultural practices—such as plant spacing and frequency and quantity of water applied. Nevertheless, when petioles test less than 500 ppm nitrate-N, the leaflets are usually a light to yellowish-green, and the petioles give a negative test for nitrate with diphenylamine reagent.

With experience, the grower can coordinate his fertilization program with the seasonal petiole nitrate-N pattern which he has found to give the best yield of quality fruit, including fruit produced with drip irrigation and slow release fertilizers. Critical values for other nutrients can be similarly used.

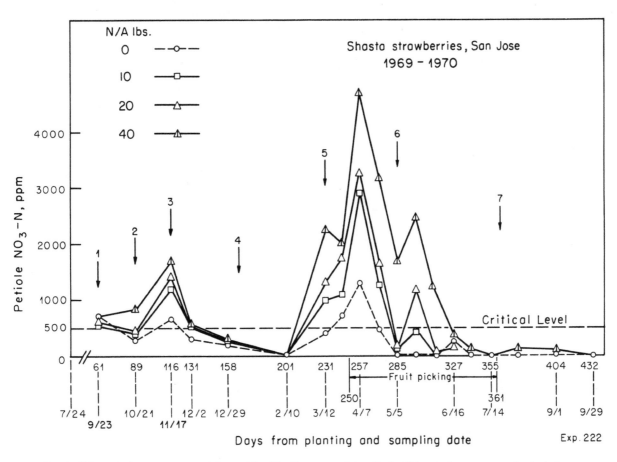

Fig. 4. Effect of nitrogen rate and time of fertilization on petiole nitrate-N levels for summer-planted strawberries treated for mite control. Leaf samples were taken just before each of the seven fertilizer treatments (indicated by arrows) were applied. Repeated application of nitrogen as small as 10 lb/acre from side-dressed ammonium nitrate can be detected by petiole analysis. These applications (with mite control) are related to yield increases (Figures 5 and 6). Nitrogen applications, indicated by numbered arrows, were not effective during late fall (3) or winter (4) and during late summer after fruit harvest (7).

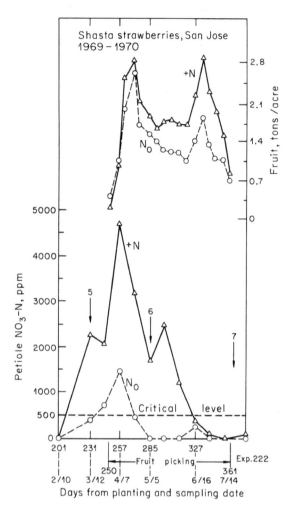

Fig. 5. Effects of nitrogen fertilization on petiole nitrate levels (bottom) and on fruit yield (top). Fruit yields are not affected appreciably until the petiole values fall below 500 ppm nitrate-N dry basis. Numbered arrows indicate side dressings of ammonium nitrate at 40 lb N/acre.

Nitrogen and yield. Specific effects of nitrogen on yield largely depend on timing and nitrogen status of the plants. This was shown in tests conducted with Shasta strawberries under field conditions in 1969-1970. The yields (see Figure 5) were not affected appreciably by nitrogen fertilization until the second crop of fruit. The petioles indicate why this happened: nitrate-N in the unfertilized plants was above the critical level until nearly all the first crop fruit had been produced. Thereafter, during the second crop, the petiole nitrate values and the yields of the unfertilized plants declined much faster than those with nitrogen. Essentially, the low petiole nitrate-N values were related to low fruit production, and conversely, high petiole nitrate-N were related to high fruit production (Table 2).

Seasonal patterns. (Figures 4, 7, and 8). In spite of differences in soil fertility and climate, in our experiments the yearly nutritional patterns for nitrogen in strawberries from the time of planting to harvest the following year were similar (see Figures 4, 7, and 8). As a rule, plants on unfertilized soil ran out of nitrogen soon after planting, even though the plants were relatively high in nitrate-N initially. Consequently, for optimum growth, nitrogen must be applied before planting or shortly after the plants have become established. On most soils, an application of 56 kg of nitrogen per hectare (50 lb N/acre) in late summer or early fall has been effective for several months after planting.

Summer-planted strawberries, as shown by plant analysis tests, became nitrogen-deficient most often during the period of dormancy in late fall and through the winter and, again, after fruit production the following summer (see Figure 4). Fertilization with nitrogen was not effective at these times, perhaps because in winter the roots were inactive in the cold soil, and in summer they became dormant physiologically after most of the crop was produced. Or, possibly, nitrogen was not effective for some other reason, although the same method of application was used at all times. Whatever the cause of failure, efficient fertilizing with nitrogen is possible only when the plants can make use of it during plant growth and fruit production.

When our plants began to grow again in the spring, nitrate-N was absorbed by the plants, and the values rose again considerably above the critical level of 500 ppm even though soil nitrogen may have been relatively low. This steep rise in petiole nitrate-N in the spring was only temporary, however, and was not sufficient to meet the nitrogen needs of the untreated plants for optimum fruit production (Figure 5 and Table 2). An early spring application of 56 kg of nitrogen per hectare (50 lb N/acre) would generally meet this need. Shortly after nitrogen fertilization and irrigation in the spring, the petiole nitrate-N value reached a maximum, followed by a rapid decline as fruit development, ripening, and further leaf growth took place. Adding nitrogen just before or after picking began delayed the decline of petiole nitrate-N during the period of heavy fruit production (see Figures 4, 5, 7, and 8). However, with summer-planted strawberries, the addition of nitrogen near the end of fruit production had little effect, and petiole nitrate-N continued to decline.

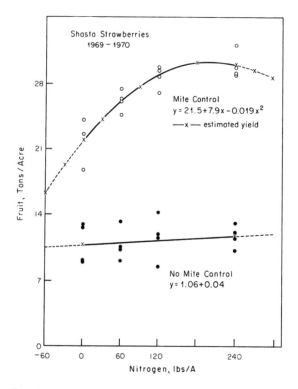

Fig. 6. Nitrogen increases fruit yield with mite control. The needs of the crop for nitrogen have been met at about 120 pounds of nitrogen per acre. However, without mite control, nitrogen fails to increase fruit yield. (See also petiole nitrate-N values in Figures 4 and 5.)

TABLE 3. Effects of Variety and Nitrogen With and Without Mite Control on Strawberry Fruit Production (San Jose, 1972, 1973, and 1974)

Fruit picking period*	Nitrogen treatment†	Mite control	Yield	
	lb/acre		tons/acre	
April 11 to			Tioga	Shasta
June 13, 1972	0	None	21.4	9.3
	0	Treated	21.2	11.2
	100	None	26.3	11.2
	100	Treated	26.6	14.3
			Tioga	Shasta
April 24 to	0	None	11.6	7.5
June 26, 1973	0	Treated	13.2	8.0
	150	None	16.8	10.8
	150	Treated	19.3	12.8
			Tioga	Tufts
April 17 to	0	None	19.9	9.3
July 15, 1974	0	Treated	23.1	15.6
	150	None	26.2	16.1
	150	Treated	27.6	24.8

†Side dressed with ammonium nitrate at the rate of 50 lb N/acre on September 7, 1971, February 29, 1972; September 12, 1972, February 20, 1973, and May 16, 1973; and September 11, 1973, January 29, 1974, and April 8, 1974.

*Fruit was picked at weekly intervals.

Nitrogen carry-over. The carry-over effect of nitrogen fertilization in the fall to the following spring varies greatly because of differences in rainfall, soil type, plant activity, and time of fertilizer application. In some instances, nitrogen carry-over can be substantial, but in most cases it will not be sufficient for maximum yields. Analyzing petioles for nitrate-N is a good way to observe the carry-over effect of previous nitrogen fertilizations and to make adjustment for it in the current fertilizer program.

Nitrogen rates. Side dressings of nitrogen as low as 11.2 kg per hectare (10 lb/acre) increased petiole nitrate-N values of deficient plants under favorable conditions of moisture, temperature, and the like (see Figure 4). Larger amounts of nitrogen produced higher concentrations of nitrate-N in the petioles, but diminishing effects on yield occurred with yearly applications above 134 kg of nitrogen per hectare (120 lb N/acre) (Figures 4 and 6).

In any event, the need for nitrogen and the effectiveness of various rates of nitrogen should be monitored by petiole nitrate-N analysis.

Varietal needs. Varieties differ in their ability to utilize nitrogen, primarily because of differences in growth habit or differences in fruit production and foraging power of the roots for soil nitrogen. For instance, Shasta, probably because of its lower foraging power for nitrate (see Figure 7), had a greater need for nitrogen than Tioga, even though Shasta leaf growth and fruit production (Table 3) were much less than for Tioga. Preliminary test results with Tufts suggested that this variety has a nitrate pattern similar to that of Shasta. Whether Tuft's greater demand for nitrate is caused by greater leaf and fruit production or by the lower nitrate-foraging power of its roots remains to be seen. In either event, the answer is with Tufts to use more nitrogen than with Tioga, but perhaps somewhat less than with Shasta. With the new leading varieties of Chandler and Selva in California, the nitrogen fertilizer required will undoubtedly differ as with the older varieties Shasta, Tioga, Tufts and Aiko, but the critical concentration of 500 ppm nitrate-N based on petioles from young mature leaves will be nearly the same.

Temperature and soil effects. There was also good evidence that Tioga roots remained active longer in the fall and became active sooner in the spring than did the roots of other varieties (see Figure 7). This means that different varieties on the same soil differed in petiole nitrate-N concentration not only because of differences in foraging power of the roots for nutrients but also because roots of some varieties are active at lower temperatures.

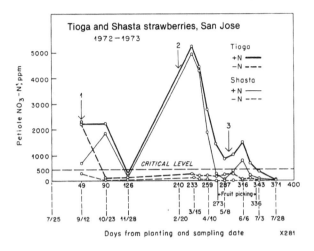

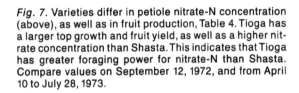

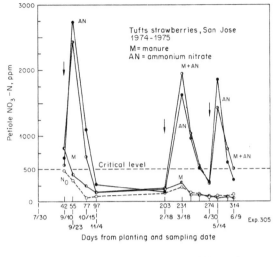

Fig. 7. Varieties differ in petiole nitrate-N concentration (above), as well as in fruit production, Table 4. Tioga has a larger top growth and fruit yield, as well as a higher nitrate concentration than Shasta. This indicates that Tioga has greater foraging power for nitrate-N than Shasta. Compare values on September 12, 1972, and from April 10 to July 28, 1973.

Fig. 8. Effect of ammonium nitrate (AN) and steer manure (M) on petiole nitrate-N for Tufts strawberries planted July 30, 1974. Steer manure (bagged, commercial grade, dried) was applied at the rate of 10 tons/acre (220 lb N/A) in the center of the planting row on July 17, 1974. Ammonium nitrate was side-dressed at the rate of 50 lb of N/A on September 10, 1974, February 18, and April 23, 1975, as indicated by arrows. Mites were controlled with Plictran®

Such temperature effects on nutrient availability should not be overlooked in the use of soil test values to indicate fertilizer needs. Obviously, the critical soil test values should be adjusted upward or downward depending not only on soil type and crop to be grown but with the strawberry variety selected. In contrast, the critical test value for petiole nitrate-N changes very little from year to year or with variety, and one can thus determine the nitrogen status of the strawberry plant anytime during the growing season and apply fertilizer nitrogen as needed.

Nitrogen Insurance. When nitrogen deficiencies have occurred in strawberries, large losses in fruit production take place (see Figure 5). In view of this, and the relatively low cost of nitrogen, the strawberry grower will often over-fertilize to make sure his crop will be getting enough nitrogen. In so doing, he may make as many as five or more side dressings of nitrogen at the rate of 50 pounds per acre. However, under most conditions only two, or possibly three, of these applications will be "seen" by the plants (see Figure 4). In our studies with summer-planted strawberries, three applications of nitrogen of 56 kg per hectare (50 lb/acre) each were sufficient—the first, shortly after the plants had become established; the second, just after

pruning, and the third, just after fruit-picking started (see Figures 7 and 8).

Manure. In a comparison of nitrogen from steer manure and from ammonium nitrate, it was found that steer manure, alone or in combination with ammonium nitrate, increased yields only from 1 to about 2 tons per acre (see Table 4), as contrasted with 7 tons per acre with ammonium nitrate alone. Furthermore, the organic nitrogen or other nutrients in the steer manure failed to enhance fruit quality, either directly or indirectly. Apparently, nitrogen from manure is not as readily available for growth as it is from ammonium nitrate, a chemical fertilizer. Petiole nitrate-N values are shown in Figure 8 and the yields in Table 4. Steer manure was applied at the recommended rate of 10 tons (which contained 220 pounds of nitrogen) per acre. The three applications of ammonium nitrate contained only 150 pounds of nitrogen per acre, but much more of it was available for vegetative growth and fruit production.

Fruit load. With most crops, a smaller fruit load lowers the need for plant nutrients, but this does not appear to be true for the nitrogen needs of strawberry. Removing the blossoms in a field experiment with 'Tioga' greatly increased leaf growth, and this caused a large decrease in petiole nitrate-N, thus indicating a need for more nitrogen. This surprising result illustrates the value of routinely determining the nitrogen status of the crop with a plant analysis program. In the aforementioned field experiment, the plant analysis results correctly indicated that the plants without fruit were nitrogen-deficient. Without a field inspection, however, one would not have known that the cause of the low petiole nitrate-N values was due to excessive leaf growth, induced by the absence of fruit. Obviously, a soil analysis alone could not have shown what effect blossom removal or blossom damage would have on the nitrogen status of the strawberry plants.

Spider mites. Plants damaged by spider mites grew poorly and, because of their poor growth, the need for nitrogen was less. Such plants have been found to be higher in nitrate-N than were comparable plants without mite damage (see Figure 9). Consequently, plants already high in nitrate-N, because of mite damage, did not respond to nitrogen fertilization. Mite control, therefore, is clearly essential if the full effects of nitrogen are to be obtained. (Figure 6).

TABLE 4. Effect of Organic and Inorganic Nitrogen on Yield and Quality of Tufts Strawberries (San Jose, 1974–1975)

Treatment	Nitrogen Per application	Nitrogen Total applied	Total yield‡	Quality Fresh	Quality Freezer
	lb/A		ton/A	percent of total yield	
None	0	0	9.3	71.0	89.2
Manure (M)*	220	220	11.1	73.0	87.4
Ammonium Nitrate (AN)†	50	150	16.5	72.7	86.1
M + AN	50	370	17.7	70.6	84.7

*Dried, commercial grade steer manure was incorporated in the beds at the rate of 10 tons per acre on July 17, 1974, 13 days before planting.

†Side dressed with ammonium nitrate at the rate of 50 lb N/acre, on September 10, 1974, February 18, 1975, and April 23, 1975.

‡Fruit was picked, weighed, and evaluated as to quality at weekly intervals from April 30 to July 1, 1975. See also figure 8.

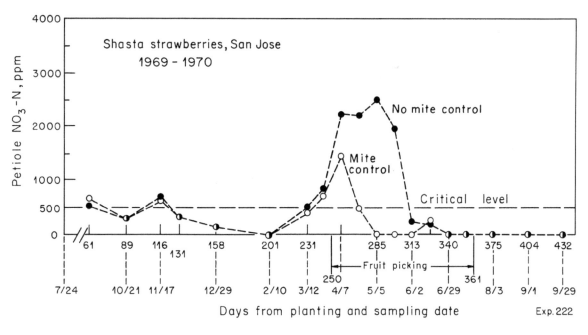

Fig. 9. Petiole nitrate-N values are higher in mite-infested plants because of reduced plant growth.

ALBINISM

Albinism, a white berry "disease" of strawberries (page 40), has occurred in nearly all strawberry-growing areas in California and in most varieties, although some varieties are more susceptible than others. Albinism frequently occurs in fields during the periods of peak fruit production that are preceded by warm weather and followed by overcast skies. Very often albino fruit appears in fields where nitrogen has been applied frequently in generous amounts and the plants have been growing vigorously. It has also appeared with heavy set of fruit associated with poor leaf development, induced by inadequate chilling of plants either before or after planting. In controlled experiments, albinism appeared with deficiencies of phosphorus, potassium, or magnesium, but not with those of nitrogen, sulfur, molybdenum, iron, manganese, zinc, boron, copper or calcium.

Low light, low sugar

From studies in plant growth chambers, low light intensity was found to produce symptoms of albinism when all other conditions were held constant (see Table 5). It appears the symptoms are directly or indirectly related to the amount of sugar produced by the strawberry plant and the partitioning of sugar for leaf growth, root development, and fruit production—with albinism occurring whenever sugar is inadequate for full fruit development. Albinism might even occur in some cultivars in high sunlight if sugar distribution fails to meet the needs of the fruit. For example, under conditions of high nitrogen, an excessive diversion of sugar to leaf growth during periods of warm, sunny weather could jeopardize fruit development, especially if this is followed by overcast, slightly warm weather. From this it follows that for best vegetative growth and fruit production, nitrogen should be kept in balance with plant needs by carefully monitoring the nitrogen supply, preferably through plant analysis.

From the standpoint of albinism, an ideal strawberry plant would be one in which the fruit has the highest priority for sugar built into its growth pattern, once the basic metabolic needs of the plant have been met. In this way the proper balance between sugar formation and its

TABLE 5. Effects of Light Intensity on Fruit Production by E-4 and Shasta Strawberries

Illumination*	Variety	Total fruit/pot		Albino fruit		Berry size	
		Weight	By count	Of total weight	Of total count	Albino	Normal
ft-c		grams	number	percent	percent	grams	grams
500	E-4	60	14.1	100	100	4.3	. . .
	Shasta	70	12.7	60	72	4.6	7.8
1,000	E-4	91	18.1	84	86	4.9	5.9
	Shasta	100	16.2	26	39	4.1	7.5
2,000	E-4	112	21.5	38	27	5.2	5.2
	Shasta	134	20.9	7	11	4.7	6.6
4,000	E-4	132	24.2	23	22	5.7	5.4
	Shasta	215	27.8	0	0	5.7	7.7

*The plants were grown in complete nutrient solution at a light intensity of 4,000 ft-c (Weston meter) for photoperiods of 14 hours and constant day and night temperature of 23°C (73°F). Thirty days after planting, when the fruit began to color, eight pots, with five plants each, were exposed to the different light intensities; all other conditions remained the same. Full sunlight on a clear day in summer is about 10,000 ft-c (foot candles).

distribution for basic metabolism, leaf growth, root development, and fruit production would be maintained. In reality, however, sugar distribution may be disrupted at any time by environmental factors, such as insufficient sunlight, improper chilling, time of planting, fertilization, irrigation, pests, and diseases; as a consequence, leaves or fruit may not develop properly or in proper sequence, and albinism would still appear.

Clearly, maintaining this delicate balance between vegetative growth and fruit production is a major concern in the development of new improved varieties. Such development will become more difficult as the pressures for larger yields and improved quality become greater. Unfortunately, as yields reach a maximum, quality suffers sooner than yield under adverse conditions. Someday, perhaps, genetic and environmental controls combined with foliar feeding of sugar and hormonal sprays will solve some of these problems. Nevertheless, just now even with the best variety, a sudden decrease in sugar production can occur because of cloudy or rainy weather, or through a large loss of leaves from pests, disease, sprays, or drought. These and other hazards must be recognized and overcome if losses from albinism are to be avoided.

HOW TO CONDUCT
SMALL-SCALE FIELD TRIALS

Corrective measures may vary with variety, soil type, climate, plant age, local cultural practices (irrigation, disease and pest control) and end use of fruit. Whenever a deficiency has been determined visually from the Color Atlas and confirmed by chemical analysis of leaves, see your farm advisor or technical consultant for the best advice on correcting the deficiency based on local conditions. However, when local experience is lacking, we recommend the following small scale experimental trial, using the amounts of material given for each element in the Color Atlas.

In the experimental trials, measure off twelve adjacent rows, 10 meters (25 feet) long. Flip a coin to determine which of the first two rows will be treated with the fertilizer. Fertilizer may be applied at planting as a side dressing close to the root zone (not at the bottom of the furrow), or as a foliar spray after the symptoms appear. Flip a coin five more times to determine the treatment "plots" in the remaining pairs of rows. Six rows, one for each pair, are thus left untreated. The experiment can be expanded to include more pairs of rows or treatments, but test results will not be valid with fewer than four pairs (eight rows). Be sure to number your plots, and record how, what, where, and when fertilization has taken place.

Roots of strawberry plants are fibrous, shallow, and grow essentially straight downward in raised beds. When the fertilizer is side-dressed within the root zone, there is very little, if any, borrowing of nutrients from adjacent plants (page 15, upper left photo) within or across the row (page 13). Placing fertilizer in the bottoms of furrows between rows leads to inefficient utilization of fertilizers by strawberry plants and possibly to objectionable leaching losses of nitrate to ground water supplies.

When amounts of material to be applied are small, an inert filler should be mixed thoroughly with the fertilizer to ensure uniformity. In the spray treatments, the addition of a detergent at 0.1 percent will improve uniformity.

At picking, record the weight of the marketable fresh fruit to the nearest gram (1/10 ounce) per row. Results for each treatment should be averaged, and when desirable, evaluated statistically, since the plot layout meets the requirements of statistical analysis of data (Little and Hills, 1978). If the corrective measures have been effective, fruit yields should have improved, concentrations of the nutrient(s) applied should have increased in the petioles or blades, and the symptoms should have disappeared. Only dust-free or cleaned leaves (washed in 0.1 N HCL and rinsed twice in distilled water) should be used for chemical analysis.

The amount of fertilizer needed per application can be calculated from the amount of the element needed per row, plot, or area, divided by the per cent of the element in the fertilizer, multiplied by 100. For example, for a material containing 20.0 per cent iron and requiring 1.0 gram iron per meter:

$$\frac{1.0\ g}{20.0\%} \times 100 = 5 \text{ grams of material per meter of row,}$$

or when 2.0 ounces of iron are required per 100 feet of row from a material containing 4 percent iron, the calculations become:

$$\frac{2.0\ oz}{4.0\ \%} \times 100 = 50 \text{ ounces per 100 feet of row.}$$

ANGLO
AND METRIC CONVERSIONS

To convert—

ounces (oz) to grams (g), multiply by 28.35.

grams to ounces, multiply by 0.035.

ounces per 100 feet (ft) of row to grams per meter (m), multiply by 1.0.

grams per meter of row to ounces per 100 feet, multiply by 1.0.

pounds (lb) per 100 gallons (gal) to grams per liter (l), multiply by 1.20.

grams per liter to pounds per 100 gallons, multiply by 0.833.

Also—

1 yard (yd) = 0.914 meter

1 meter = 1.094 yards or 3.282 feet

1 inch (in) = 2.54 centimeters (cm)

1 centimeter = 0.394 inch

1 pound = 454 grams

1 kilogram (kg) = 2.205 pounds

1 pound per acre = 1.12 kilogram per hectare (ha)

1 kilogram per hectare = 0.891 pound per acre

1 acre = 43,560 square feet

1 acre = 0.405 hectare

1 hectare = 2.471 acres

1 quart (qt) = 0.946 liter

1 liter = 1.057 quarts

1 percent (%) = 10,000 parts per million (ppm)

10,000 parts per million = 1 percent

1 microgram per gram (1.0 μg^{-1}) = 1.0 parts per million

1.0 part per million = 1 microgram per gram

GENERAL REFERENCES

Anderson, W. 1969. *The strawberry: a world bibliography 1920-1966.* New Jersey: Scarecrow Press, Inc. 731 pp.

Bringhurst, R. S., and D. A. Khan. 1963. Natural pentaploid *F. chiloensis—F. vesca* hybrids in coastal California and their significance in polyploid *Frageria* evolution. *Am. J. Bot.* 50:658-61.

California Fertilizer Association. 1985. *Western fertilizer handbook,* 7th ed. Sacramento. 288 pp.

Darrow, G. M. 1966. *The strawberry: history, breeding, and physiology.* New York: Holt, Reinhart, and Winston. 447 pp.

Johnson, C. M., and A. Ulrich. 1959. Analytical methods for use in plant analysis. *Univ. of Calif. Agric. Exp. Sta. Bull.* 766: pp. 26-78.

Little, T. M., and F. J. Hills. 1978. *Agricultural experimentation, design, and analysis.* New York: John Wiley and Sons. 350 pp.

Otterbacher, A. G., and R. M. Skirvin. 1978. Derivation of the binomial *Frageria x ananassa* for the cultivated strawberry. *Hort. Sci.* 13:637-39.

Roudeillac, M. P. et al. 1987. *La Fraise: techniques de production / realisation.* Paris: Centre technique interprofesionel des fruits et legumes. 384 pp.

Schrader, W. L., and N. C. Welch. 1990. Salinity management in strawberry production. *Calif. Strawberry News Bull.* Dec. 10, 1990. 11 pp.

Sprague, H. B. 1964. *Hunger signs in crops.* 3rd ed. New York: David McKay Co., Inc. 461 pp.

Thomas, H. E., and E. V. Goldsmith. 1945. The Shasta, Sierra, Lassen, Tahoe, and Donner strawberries. *Univ. of Calif. Agric. Exp. Sta. Bull.* 690. 12 pp.

Ulrich, A., and F. J. Hills. 1969. *Sugarbeet nutrient deficiency symptoms: a color atlas and chemical guide.* Publ. 4051. Oakland: Univ. of Calif. Div. of Agric. Sci. 40 pp.

Welch, N. 1989. *Strawberry production in California.* Leaflet 2959. Oakland: Univ. of Calif. Div. of Agric. and Nat. Resources. 16 pp.

Wilhelm, S., and J. E. Sagen. 1974. *A history of the strawberry from ancient gardens to modern markets.* Publ. 4031. Oakland: Univ. of Calif. Div. of Agric. Sci. 298 pp.

ACKNOWLEDGMENTS

The authors greatly appreciate the technical suggestions from Robert S. Ayers, the organizational and editorial aid from Peggy Anne Davis, and the encouragement and advice from Harold E. Thomas, Stephen Wilhelm, Royce S. Bringhurst, and Harold Johnson. Others who have our thanks include Sheryl Liebscher for the design and production coordination of this publication, Frank Murillo for the preparation of the figures, Clifford Carlson for many photographs of plant symptoms and culturing plants in the greenhouse, Kwok Fong and Carlos J. Llano for numerous analyses of plant tissues, and Pamela L. Colville and Thomas M. Kretchun for their assistance in the field. We acknowledge and thank the California Strawberry Advisory Board whose financial support made this research project possible.

Many persons made it possible for the senior author to complete this publication. For her special efforts, he dedicates his work most warmly to his wife, Jane M. Ulrich, and to others who also helped make his recovery possible: Dr. Chariklia T. Spiegal, a dear friend and medical advisor at George Washington University Medical Center; Dr. William Raskoff, Kaiser Permanente Foundation in San Francisco, and Dr. Norman Shumway, his associates and the nursing staff, Stanford University.

CPSIA information can be obtained
at www.ICGtesting.com
Printed in the USA
BVHW02n0022270818
525651BV00012B/61/P